Collins

11+
Maths

Quick Practice Tests
Ages 10-11
Book 2

Anne Stothers

Contents

ACKNOWLEDGEMENTS

The author and publisher are grateful to the copyright holders for permission to use quoted materials and images.

Every effort has been made to trace copyright holders and obtain their permission for the use of copyright material. The author and publisher will gladly receive information enabling them to rectify any error or omission in subsequent editions. All facts are correct at time of going to press.

Published by Collins
An imprint of HarperCollins*Publishers* Limited
1 London Bridge Street
London SE1 9GF

HarperCollins*Publishers*
Macken House, 39/40 Mayor Street Upper,
Dublin 1, D01 C9W8, Ireland

ISBN: 978-0-00-870119-2
First published 2025

10 9 8 7 6 5 4 3 2 1

© HarperCollins*Publishers* Limited 2025

British Library Cataloguing in Publication Data.

A CIP record of this book is available from the British Library.

Author: Anne Stothers
Publisher: Clare Souza
Project Manager and Editor: Richard Toms
Cover Design: Sarah Duxbury
Text and Page Design: Ian Wrigley
Layout and Artwork: Q2A Media
Production: Bethany Brohm

Printed in India by Multivista Global Pvt. Ltd.

MIX
Paper | Supporting
responsible forestry
FSC™ C007454

About this book

Familiarisation with 11+ test-style questions is a critical step in preparing your child for the 11+ selection tests. This book gives children lots of opportunities to test themselves in short, manageable bursts, helping to build confidence and improve the chance of test success.

It contains 22 tests designed to develop key numeracy skills. An example question and answer can be found at the start of Test 1.

- Each test is designed to be completed within a short amount of time. Frequent, short bursts of revision are found to be more productive than lengthier sessions.

- GL Assessment tests can be quite time-pressured so these practice tests will help your child become accustomed to this style of questioning.

- We recommend your child uses a pencil to complete the tests, so that they can rub out the answers and try again at a later date if necessary.

- Children will need a pencil and a rubber to complete the tests as well as some spare paper for rough working. They will also need to be able to see a clock/watch and should have a quiet place in which to do the tests.

- Your child should **not** use a calculator for any of these tests.

- Answers to every question are provided at the back of the book, with explanations given where appropriate.

- After completing the tests, children should revisit their weaker areas and attempt to improve their scores and timings.

For more information about 11+ preparation and other practice resources available from Collins, go to our website at:

collins.co.uk/11plus

Test 1

You have 10 minutes to complete this test.

You have 10 questions to complete within the given time.

Circle the letter above the correct answer.

EXAMPLE

Which number should go in the box?

$$\frac{4}{7} = \frac{12}{\square}$$

A	B	Ⓒ	D	E
7	14	21	28	48

① What is the value of the 3 in the number 42.032?

A	B	C	D	E
3 thousandths	3 hundredths	3 tenths	3 ones	3 tens

② What is the product of 18 and 5?

A	B	C	D	E
45	50	54	80	90

③ An aeroplane leaves London at 22:45 and lands in Cape Town the next morning at 09:37.

How long was the flight?

A	B	C	D	E
10 hours 52 minutes	9 hours 52 minutes	11 hours 25 minutes	9 hours 57 minutes	10 hours 7 minutes

(4) Work out the value of −3 − 8 + 6

A	B	C	D	E
−1	−5	−17	11	1

(5) Three cakes and four teas cost £13.75

Five cakes cost £11.25

How much does one tea cost?

A	B	C	D	E
£2.25	£7.00	£4.75	£1.75	£2.50

(6) Sam has five pieces of wood.

The lengths of the pieces are:

    1.1 m   67 cm   200 mm   56 cm   1.2 m

Calculate the mean length of the three longest pieces of wood.

A	B	C	D	E
99 cm	1 m	900 mm	0.9 m	97 cm

(7) The diagram below shows some angles.

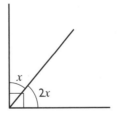

What is the value of x?

A	B	C	D	E
90°	180°	60°	45°	30°

Questions continue on next page

(8) Here is a sequence of patterns made using oval shapes.

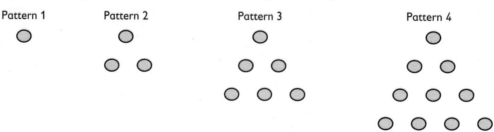

Pattern 1   Pattern 2   Pattern 3   Pattern 4

How many shapes will be in the sixth pattern?

A	B	C	D	E
14	21	15	18	22

(9) The bar chart shows the number of pounds raised by five children for charity.

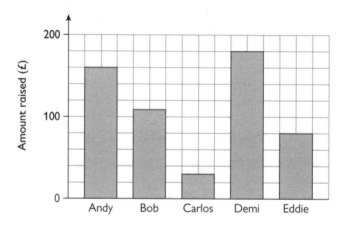

What is the difference between the total amount raised and the amount that Demi raised?

A	B	C	D	E
£560	£460	£380	£360	£470

(10) Tom thinks of a number, doubles it, and then adds 13.

The answer is 27.

What is the number Tom first thought of?

A	B	C	D	E
7	20	40	14	15

Score: .... / 10

Test 2

You have 10 minutes to complete this test.

You have 10 questions to complete within the given time.

Circle the letter above the correct answer.

(1) You are given that 237 × 64 = 15 168

What is 237 × 128?

A	B	C	D	E
15 168	30 336	7584	75 840	474

(2) A bag contains 24 marbles.

$\frac{1}{4}$ of the marbles are blue.

The rest of the marbles are green.

How many green marbles are there?

A	B	C	D	E
18	6	16	8	21

(3) One day, the temperature in Toronto was 20°C lower than in London.

If the temperature in London was 13°C, what was the temperature in Toronto?

A	B	C	D	E
−20°C	−13°C	7°C	−3°C	−7°C

Questions continue on next page

④ This shape is made from six squares.

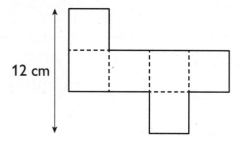

12 cm

Calculate the perimeter of the shape.

A	B	C	D	E
24 cm	96 cm	48 cm	56 cm	84 cm

⑤ Which of the following shapes is a pentagon?

A	B	C	D	E

⑥ A sequence starts 3, 4, 6, 9, 13, ...

What is the next term in the sequence?

A	B	C	D	E
14	15	16	17	18

⑦ Raj buys seven pens.

Each pen costs 83p.

He pays with a £10 note.

How much change should Raj get?

A	B	C	D	E
£3.29	£9.17	£4.19	£4.12	£4.81

(8) What number is 45 000 smaller than six million?

A	B	C	D	E
5 955 000	5 945 000	5 999 055	5 995 500	5 550 000

(9) n stands for a whole number.

5 lots of n are equal to 30.

What are 3 lots of n equal to?

A	B	C	D	E
18	10	6	90	15

(10) Twenty-five lamp posts are spaced evenly along a street that is 840 metres long.

What is the distance between each pair of lamp posts?

A	B	C	D	E
36 metres	30 metres	33 metres	35 metres	32 metres

Test 3

You have 10 minutes to complete this test.

You have 10 questions to complete within the given time.

Circle the letter above the correct answer.

1. Billy drives his truck 423 kilometres every day for 14 days.

   How many kilometres does Billy drive in total?

A	B	C	D	E
5902	5892	5922	5822	5912

2. In a pet shop, there are three times as many female hamsters as male hamsters.

   There are 36 hamsters altogether.

   How many male hamsters are there in the pet shop?

A	B	C	D	E
27	12	24	108	9

3. Six children complete a timed puzzle.

   Their start and finish times are recorded in the table.

Name	Start time	Finish time
Adam	10.00 am	10.19 am
Bev	10.15 am	10.46 am
Carlo	10.30 am	11.01 am
Dev	10.55 am	11.22 am
Emir	11.05 am	11.31 am
Fran	11.25 am	11.44 am

   Work out the total time that the children took to complete the puzzle.

A	B	C	D	E
150 minutes	151 minutes	152 minutes	153 minutes	154 minutes

(4) Remi is laying a carpet in a room.

The diagram shows a plan of the room.

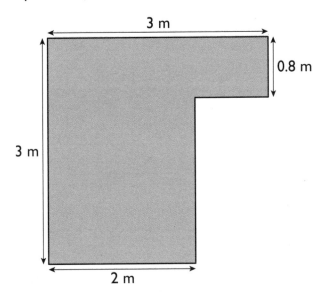

How much carpet does Remi need?

A	B	C	D	E
6.8 m²	6 m²	3.8 m²	9 m²	9.8 m²

(5) A jug contains 3.5 litres of water.

Fay makes a drink by using one-fifth of the water in the jug.

Paul then makes a drink using 650 ml of water from the jug.

How much water is left in the jug?

A	B	C	D	E
2150 ml	2450 ml	2.5 litres	2.35 litres	2800 ml

(6) Dan buys a phone which costs £850.

He pays a deposit of £275 and then he makes 23 equal monthly payments.

How much is each monthly payment?

A	B	C	D	E
£35	£25	£15	£30	£20

Questions continue on next page

⑦ Work out 4.73 ÷ 100

A	B	C	D	E
473	0.00473	0.0473	47.3	0.473

⑧ What is the value of x in the equation below?

$4x - 8 = 32$

A	B	C	D	E
2	4	6	8	10

⑨ The diagram shows an equilateral triangle.

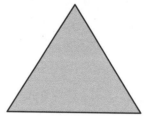

How many lines of symmetry does the equilateral triangle have?

A	B	C	D	E
0	1	2	3	4

⑩ Work out the mean of the three fractions.

$$\frac{1}{2} \quad \frac{2}{3} \quad \frac{3}{4}$$

A	B	C	D	E
$\frac{23}{12}$	$\frac{23}{36}$	$\frac{2}{9}$	$\frac{2}{3}$	$\frac{6}{24}$

Score: .... / 10

Test 4

You have 10 minutes to complete this test.

You have 10 questions to complete within the given time.

Circle the letter above the correct answer.

1 What is the value of the 9 in 8 965 471?

A	B	C	D	E
9 million	9 hundred thousand	90 thousand	9 thousand	9 hundred

2 The pictogram shows information about the number of phones sold by a shop in March, April, May and June.

March	
April	
May	
June	

Key:

represents 8 phones

What was the total number of phones sold in these four months?

A	B	C	D	E
33	66	34	64	56

3 What is 0.12 as a fraction in its lowest terms?

A	B	C	D	E
$\frac{1.2}{10}$	$\frac{12}{10}$	$\frac{12}{100}$	$\frac{3}{25}$	$\frac{6}{50}$

Questions continue on next page

(4) $215 \div \boxed{} = 43$

What number does $\boxed{}$ stand for?

A	B	C	D	E
7	4	6	15	5

(5) A woman was 26 years old when her daughter was born.

Now she is three times as old as her daughter.

How old is her daughter now?

A	B	C	D	E
14	39	13	52	15

(6) Jo started watching a video at 7.56 pm.

The video was 1 hours 53 minutes long.

Jo paused the video halfway through for 19 minutes to chat to her friend.

At what time did Jo finish watching the video?

A	B	C	D	E
10.19 pm	10.08 pm	9.08 pm	10.12 pm	11.08 pm

(7) $b - 7 = 5a$

Which of the following is **not** true?

A	B	C	D	E
$7 = b - 5a$	$b = 5a + 7$	$b = 5a - 7$	$a = \dfrac{b-7}{5}$	$b = 7 + 5a$

(8) Purple paint is made by mixing 1 part white paint to 2 parts red paint to 2 parts blue paint.

How many litres of red paint are needed to make 25 litres of purple paint?

A	B	C	D	E
10 litres	5 litres	20 litres	15 litres	8 litres

How many small cubes can be fitted into the large cube?

2 cm

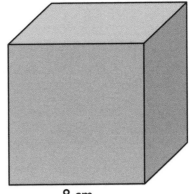

8 cm

A	B	C	D	E
64	8	16	32	128

10 The table shows the values of some shapes added together.

🌰	⚛	⚛	17
🌰	🌰	⚛	19
▱	▱	🌰	11

What is the value of ▱ ?

A	B	C	D	E
5	2	3	7	4

Score: .... / 10

15

Test 5

You have **10 minutes** to complete this test.

You have **10 questions** to complete within the given time.

Circle the letter above the correct answer.

① What is this number in digits?

Four hundred and five thousand and twenty-seven

A	B	C	D	E
450 027	400 527	405 027	452 700	40 527

② A school has 1037 students.

215 new students join the school and 178 students leave the school.

How many students are now at the school?

A	B	C	D	E
1252	1074	1000	1072	1254

③ A plant was 93 centimetres tall.

It grows by another 29 centimetres.

How tall is the plant now?

A	B	C	D	E
1.12 m	122 m	112 m	1.22 m	11.2 m

(4) Which of these expressions represents the size of angle x?

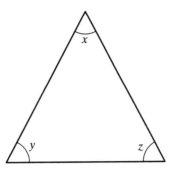

A	B	C	D	E
$90° - (y + z)$	$360° - (y + z)$	$90° + y + z$	$(y + z) - 180°$	$180° - (y + z)$

(5) A number is multiplied by 4 and then 9 is subtracted.

The answer is 27.

What is the number?

A	B	C	D	E
4	5	7	9	36

(6) Which number should go in the box to make this calculation correct?

$$2.73 ÷ \boxed{} = 273$$

A	B	C	D	E
0.1	0.01	0.001	101	100

(7) Annie is travelling from Leicester to Manchester by train.

Her journey involves taking a 66-minute train to Sheffield, followed by a 23-minute wait.

She then takes a 54-minute train to Manchester.

If her train leaves Leicester at 08:49, at what time will she arrive in Manchester?

A	B	C	D	E
11:12	10:32	12:22	11:22	12:12

Questions continue on next page

8 The area of the triangle is 56 cm².

What is the value of x?

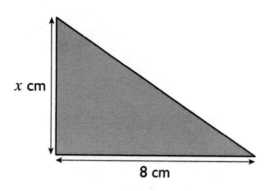

x cm

8 cm

A	B	C	D	E
7	8	6	14	12

9 Which net **cannot** be folded into a cube?

A	B	C	D	E

10 What is the difference between the largest and the smallest prime numbers between 20 and 30?

A	B	C	D	E
8	6	5	4	2

Score: .... / 10

Test 6

You have 10 minutes to complete this test.

You have 10 questions to complete within the given time.

Circle the letter above the correct answer.

1. A bookcase has 5 shelves.

   Each shelf can hold 19 books.

   How many books can the bookcase hold?

A	B	C	D	E
55	115	145	95	105

2. If 26 × 18 = 468, what is the answer to 46.8 ÷ 1.8?

A	B	C	D	E
26	2.6	260	0.26	20.6

3. What is the missing number in this sequence?

   765 ☐ 731 714 697

A	B	C	D	E
749	748	743	747	746

Questions continue on next page

(4) The chart below shows how far Jay ran on five days.

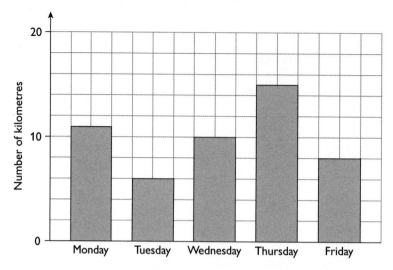

How many kilometres did he run altogether?

A	B	C	D	E
5	52	58	59	50

(5) Work out $4^2 - 3 \times 5$

A	B	C	D	E
1	65	5	25	−7

(6) Find the difference between $\dfrac{1}{12} + \dfrac{1}{4}$ and $\dfrac{1}{3} \div 2$

A	B	C	D	E
$\dfrac{1}{4}$	$\dfrac{1}{6}$	$\dfrac{1}{3}$	$\dfrac{1}{12}$	$\dfrac{1}{8}$

(7) The rectangle shown on the grid is translated so that point A is now at the coordinates (3, 8).

What are the coordinates of point C after the translation?

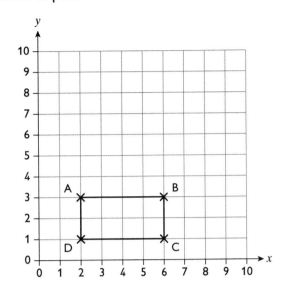

A	B	C	D	E
(7, 8)	(7, 6)	(7, 7)	(3, 6)	(6, 7)

(8) What is 40% of $\frac{5}{8}$ of 240?

A	B	C	D	E
60	96	12	150	48

(9) Three coffees and four teas cost £15.45

Five coffees cost £11.75

How much does one tea cost?

A	B	C	D	E
£8.40	£2.35	£4.20	£2.10	£2.40

(10) If a clock shows 2:30, what is the size of the smaller angle between the hour and minute hands?

A	B	C	D	E
90°	120°	115°	110°	105°

Score: .... / 10

Test 7

You have 10 minutes to complete this test.

You have 10 questions to complete within the given time.

Circle the letter above the correct answer.

(1) What is 67.892 rounded to the nearest tenth?

A	B	C	D	E
68.0	67.89	67.8	67.9	70

(2) Lucy earns £9.50 per hour.

She works for 8 hours per day and 5 days per week.

How much does she earn in three weeks?

A	B	C	D	E
£142.50	£1140	£380	£228	£1596

(3) Work out the answer to 126 × 37

A	B	C	D	E
4622	3642	4042	3618	4662

(4) Convert 1 000 000 mm to kilometres.

A	B	C	D	E
0.1 km	1 km	10 km	100 km	1000 km

(5) The first five terms of a sequence are 4, 11, 18, 25, 32

Find the 11th term of the sequence.

A	B	C	D	E
74	70	68	76	72

(6) Bill and Ben share some money in the ratio 5 : 3

Ben gets £24

How much does Bill get?

A	B	C	D	E
£64	£24	£40	£48	£4.80

(7) Work out 16% of 450

A	B	C	D	E
45	67.5	63	270	72

(8) How many small rectangles will fit into the large rectangle?

3 cm

1 cm

4 cm

15 cm

A	B	C	D	E
15	16	17	20	18

Questions continue on next page

(9) Which of these statements is **not** true for this regular octagon?

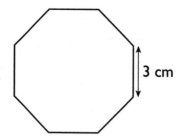

3 cm

A	B	C	D	E
There are eight equal sides.	There are eight equal angles.	The perimeter is 24 cm.	There are eight lines of symmetry.	There is only one pair of parallel sides.

(10) The sum of three consecutive odd numbers is 45

What is the **smallest** of the three numbers?

A	B	C	D	E
23	13	21	15	17

Score: .... / 10

Test 8

You have 10 minutes to complete this test.

You have 10 questions to complete within the given time.

Circle the letter above the correct answer.

1 What is 2.9 as a fraction?

A	B	C	D	E
$\frac{29}{10}$	$\frac{2.9}{10}$	$\frac{29}{100}$	$\frac{10}{29}$	$\frac{290}{10}$

2 The pictogram shows the number of pizzas sold by a shop last week.

Monday	◐
Tuesday	◐ ◖
Wednesday	◖
Thursday	◐ ◐ ◖
Friday	◐ ◐ ◐ ◐

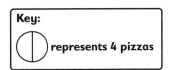

Key:

◐ represents 4 pizzas

What is the total number of pizzas sold by the shop last week?

A	B	C	D	E
21	38	12	44	40

3 To find the sum of the angles inside a polygon, you can subtract 2 from the number of sides and multiply this by 180.

A decagon has 10 sides.

What is the sum of the angles inside a decagon?

A	B	C	D	E
1800°	1620°	1440°	1080°	1260°

Questions continue on next page

④ The end points of five lines are shown in the answer options below.

Which line is parallel to the line in the diagram?

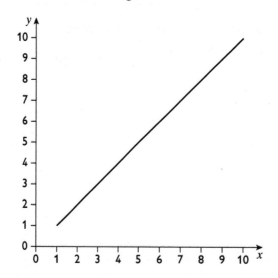

A	B	C	D	E
(1, 1) and (10, 1)	(10, 10) and (1, 9)	(2, 1) and (3, 9)	(1, 4) and (7, 10)	(0, 0) and (9, 10)

⑤ What is the area of this triangle?

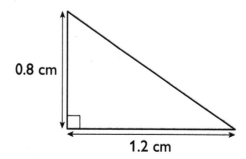

0.8 cm

1.2 cm

A	B	C	D	E
0.48 cm²	0.96 cm²	4.8 cm²	9.6 cm²	0.24 cm²

⑥ In a sale, a £125 table is reduced in price by 20%.

It is then reduced by a further 5%.

What is the cost of the table after the reductions?

A	B	C	D	E
£100	£93.75	£90	£95	£85

(7) To change from kilometres into miles, you can divide the number of kilometres by 8 and then multiply by 5.

Which conversion is **not** correct?

A	B	C	D	E
104 kilometres = 65 miles	56 kilometres = 30 miles	888 kilometres = 555 miles	40 kilometres = 25 miles	160 kilometres = 100 miles

(8) The perimeter of an equilateral triangle is 22.5 cm.

What is the length of each side?

A	B	C	D	E
7 cm	7.5 cm	5.625 cm	11.25 cm	6.5 cm

(9) Helen goes to a café.

She buys:

    2 coffees at £2.65 each

    3 teas at £2.45 each

    1 hot chocolate at £2.95

Work out the total amount that Helen spends.

A	B	C	D	E
£16.20	£13.15	£16.10	£12.95	£15.60

(10) $2^3 \times 2^2 \times 2$ is **not** the same as which of the following?

A	B	C	D	E
$2^3 \times 2^3$	4^3	$4^2 \times 4$	$2^3 \times 4$	8^2

Test 9

You have 10 minutes to complete this test.

You have 10 questions to complete within the given time.

Circle the letter above the correct answer.

(1) Work out $(2 × -4) - (-2 × 2)$

A	B	C	D	E
−12	12	−4	4	−8

(2) Which number should go in the box to make the number sentence correct?

$73 × 98 = 7300 - \boxed{}$

A	B	C	D	E
73	146	98	196	171

(3) Amy drives from London to Manchester.

She leaves London at 8.45 am.

Amy drives for $2\frac{3}{4}$ hours before stopping for a 20-minute break.

She then drives for another 95 minutes.

At what time does Amy arrive in Manchester?

A	B	C	D	E
1.25 pm	12.25 pm	1.05 pm	12.20 pm	12.55 pm

(4) The length of each side of an equilateral triangle is $(x + 7)$ cm.

The perimeter of the equilateral triangle is 28.5 cm.

Work out the value of x.

A	B	C	D	E
4.5	2.5	4	9.5	0.125

(5) Which of these is **not** a factor of 36?

A	B	C	D	E
12	9	6	3	7

(6) This graph converts British pounds (£) into euros (€).

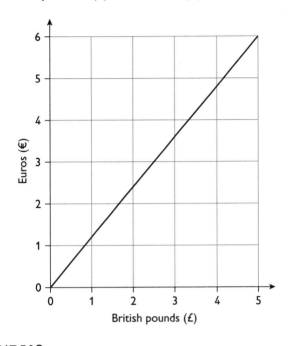

How many euros is £17.50?

A	B	C	D	E
€18	€21.50	€20.50	€21	€22

(7) Two bowls have the following capacities.

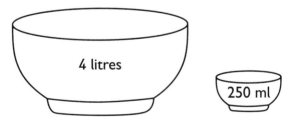

How many times would you have to fill the 250 ml bowl to make 4 litres?

A	B	C	D	E
40	16	4	8	12

Questions continue on next page

(8) Here is an equilateral triangle:

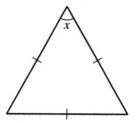

What is the size of angle x?

A	B	C	D	E
45°	90°	30°	60°	180°

(9) In one week, the hens on a farm laid 365 eggs.

The eggs were put into boxes of 6.

How many boxes were filled?

A	B	C	D	E
600	61	66	6	60

(10) A cricket club has three different memberships: senior, adult and junior.

$\frac{1}{4}$ of the members are seniors.

$\frac{3}{5}$ of the members are adults.

87 members are juniors.

How many members does the cricket club have altogether?

A	B	C	D	E
580	130	131	435	740

Score: .... / 10

Test 10

You have 10 minutes to complete this test.

You have 10 questions to complete within the given time.

Circle the letter above the correct answer.

1 What is the product of 0.1 and 0.3?

A	B	C	D	E
3	0.3	0.03	0.003	0.0003

2 The cost of one cake is £1.65

A shop is offering a special deal of 12 cakes for £15

I want to buy 12 cakes.

How much will I save by using the special deal?

A	B	C	D	E
£1.50	£3.15	£3.60	£4.80	£5.00

3 Josie has an appointment at 11.10 am.

She arrives at 10.57 am.

How minutes early is she?

A	B	C	D	E
10 minutes	17 minutes	7 minutes	3 minutes	13 minutes

Questions continue on next page

④ The chart shows the number of hours of sunshine recorded in a town over five days.

There were a total of 26 hours of sunshine recorded over the five days.

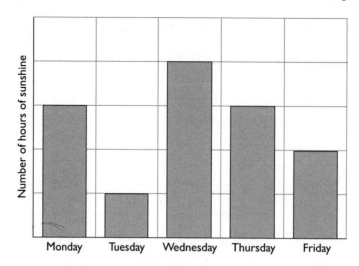

On which day were there 4 hours of sunshine?

A	B	C	D	E
Monday	Tuesday	Wednesday	Thursday	Friday

⑤ Here is a diagram of a triangle.

Each side is labelled using an algebraic expression.

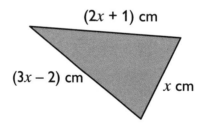

If $x = 2$, what is the perimeter of the triangle?

A	B	C	D	E
5 cm	12 cm	11 cm	6 cm	15 cm

⑥ $-3 < x < 2$

Which of the following is **not** a possible value of x?

A	B	C	D	E
−3	−2	−1	0	1

(7) Here are some shapes on a grid.

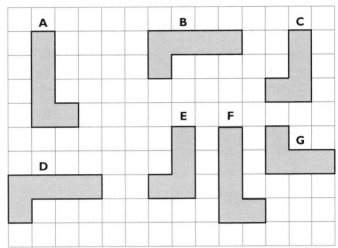

Which one of these moves is **not** a translation?

A	B	C	D	E
A to F	C to G	B to D	C to E	D to B

(8) What percentage of £7 is 35p?

A	B	C	D	E
5%	20%	3%	2%	50%

(9) The probability that it will rain tomorrow is 0.43

What is the probability that it will **not** rain tomorrow?

A	B	C	D	E
0.5	0.43	0.47	0.57	0.67

(10) Bill is four times older than I am now.

Five years ago, Bill was five times older than I was at that time.

How old am I now?

A	B	C	D	E
10	15	20	25	30

Score: .... / 10

Test 11

You have 10 minutes to complete this test.

You have 10 questions to complete within the given time.

Circle the letter above the correct answer.

① Which number correctly completes this calculation?

$7.2 \div \boxed{} = 720$

A	B	C	D	E
1000	0.01	100	0.1	10

② The lengths of five pieces of string are:

20 cm   24 mm   1.2 m   0.1 m   0.03 m

What is the difference between the two smallest lengths of string?

A	B	C	D	E
4 mm	6 mm	4 cm	6 cm	0.6 m

③ What is the sixth prime number?

A	B	C	D	E
9	12	17	11	13

④ 24 students travel by train from Leicester to London and back (return).

The total ticket price is £625.20

How much does each student pay for their return train ticket?

A	B	C	D	E
£25.06	£20.65	£26.05	£20.56	£26.50

5 Three pens cost £2.37

Two rulers cost £1.12

Anju wants to buy 10 pens and 5 rulers.

How much does this cost Anju in total?

A	B	C	D	E
£10.70	£3.39	£9.91	£13.50	£9.70

6 Beans are sold in two sizes of tins.

Rose buys three large tins of beans and some small tins of beans.

In total, she buys 3500 g of beans.

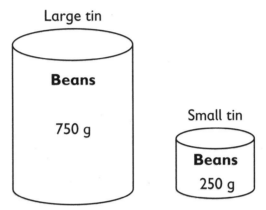

Large tin

Beans

750 g

Small tin

Beans

250 g

Work out the number of small tins of beans that Rose buys.

A	B	C	D	E
3	4	7	6	5

7 Look at this angle.

x

Which statement is correct?

A	B	C	D	E
Angle $x < 90°$	Angle $x = 180°$	Angle $x < 180°$	Angle $x > 180°$	Angle x is an obtuse angle

Questions continue on next page

(8) Here are four number cards:

A fifth card is added and the mean of all five cards is 6.

What is the value on the fifth card?

A	B	C	D	E
3	4	5	6	7

(9) Which number is 0.007 greater than 6.098?

A	B	C	D	E
6.105	6.168	6.798	6.195	6.095

(10)

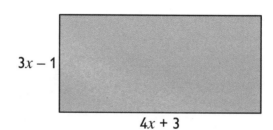

$3x - 1$

$4x + 3$

Which expression does **not** represent the perimeter of this rectangle?

A	B	C	D	E
$7x + 2$	$14x + 4$	$7x + 2 + 7x + 2$	$2(7x + 2)$	$2(3x - 1) + 2(4x + 3)$

Test 12

Circle the letter above the correct answer.

1 Four buses can carry 57 passengers each.

How many passengers in total can the buses carry?

A	B	C	D	E
221	214	228	208	288

2 Work out $\frac{2}{3} - \frac{1}{4}$

A	B	C	D	E
$\frac{7}{12}$	$\frac{1}{-1}$	$\frac{1}{12}$	$\frac{5}{12}$	$\frac{1}{3}$

3 The length of a road is 13 457 metres.

Convert the length of the road from metres into kilometres.

A	B	C	D	E
1345.7 km	134.57 km	13.457 km	1.3457 km	13 457 000 km

4 1, 2, 3, 5, 7

Which of these numbers is **not** a prime number?

A	B	C	D	E
1	2	3	5	7

Questions continue on next page

(5) The pictogram shows the number of ice creams sold by a shop last week.

Monday	◯
Tuesday	◯ ◯ ◯
Wednesday	◯ ◯

A total of 32 ice creams were sold on Monday and Tuesday.

How many ice creams were sold on Wednesday?

A	B	C	D	E
8	4	2	10	16

(6) Work out 7% of £70

A	B	C	D	E
£49	£10	£7	£4.90	£7.70

(7) Pooja has two dogs.

Each dog eats $\frac{3}{4}$ of a tin of dog food each day.

How many tins of dog food will Pooja need to feed her dogs for one week?

A	B	C	D	E
10	11	12	13	14

(8) A farmer has sheep and chickens in the ratio 1 : 4

If the farmer has 100 chickens, how many sheep do they have?

A	B	C	D	E
20	80	25	125	100

(9) Here are five shapes:

  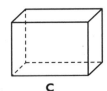 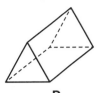

A   B   C   D   E

Which shape is **not** a prism?

A	B	C	D	E
A	B	C	D	E

(10) Alex is x years old.

Alex's sister, Beth, is two years older than him.

Marie is twice as old as Beth.

What is the correct expression for Marie's age in terms of x?

A	B	C	D	E
$2(x - 2)$	$2x - 2$	$2(x + 2)$	$2x + 2$	$2x$

Test 13

You have **10** minutes to complete this test.

You have **10** questions to complete within the given time.

Circle the letter above the correct answer.

① How many times greater is the value of the digit 3 in 3 429 025 than the value of the digit 3 in 8 036 504?

A	B	C	D	E
10	100	1000	10 000	100 000

② Abby and Ben saved £141.

Abby saved £27 more than Ben.

How much did Ben save?

A	B	C	D	E
£57	£141	£84	£114	£168

③ A marble costs 47p.

Beth has £5 and buys as many marbles as she can.

Work out the amount of change Beth should get from £5.

A	B	C	D	E
47p	60p	17p	30p	3p

④ What is $\frac{3}{4}$ of $\frac{1}{3}$?

A	B	C	D	E
$\frac{4}{12}$	$\frac{1}{3}$	$\frac{4}{7}$	$\frac{3}{4}$	$\frac{1}{4}$

(5) Katya's plant is 1.03 metres tall.

It has grown 17 centimetres in the last month.

How tall was the plant last month, in metres?

A	B	C	D	E
0.96 m	0.86 m	0.94 m	96 m	86 m

(6) Shilpa has three different boxes. The mass of each of them is shown.

What is the total mass of these boxes in kilograms?

A	B	C	D	E
293 kg	2930 kg	2.93 kg	1.545 kg	1671 kg

(7) 1.6 kilometres equal 1 mile.

How many kilometres equal 5 miles?

A	B	C	D	E
8	7	6	5	4

(8) The second and fourth terms in a sequence are −1 and 9.

What is the sixth term in the sequence?

A	B	C	D	E
49	23	24	19	41

Questions continue on next page

⑨ The graph shows Jakub's journey to his friend's house.

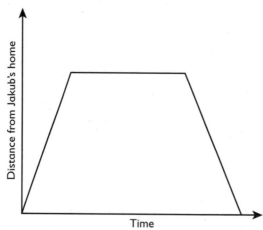

Which statement best describes Jakub's journey?

A	B	C	D	E
Jakub cycled to his friend's house, and then cycled straight back.	Jakub cycled to his friend's house, stayed for lunch and then cycled back, stopping on the way to buy water.	Jakub cycled back from his friend's house.	Jakub cycled to his friend's house, stayed for lunch, and then cycled back.	Jakub was at his friend's house, stayed for lunch, and then cycled back.

⑩ What is 2.7 ÷ 30?

A	B	C	D	E
9	90	0.9	0.009	0.09

Score: .... / 10

Test 14

You have 10 minutes to complete this test.

You have 10 questions to complete within the given time.

Circle the letter above the correct answer.

(1) Jade has 47 pens.

She sorts them into three pencil cases.

Each pencil case can hold 13 pens.

How many pens don't fit into Jade's pencil cases?

A	B	C	D	E
6	7	8	9	10

(2) Which of these is a multiple of 7?

A	B	C	D	E
1	12	23	37	42

(3) What is the difference between 40% of 140 and 25% of 220?

A	B	C	D	E
1	2	3	4	5

(4) 3.5 kg of potatoes cost £3.15

What is the cost of 2.5 kg of potatoes?

A	B	C	D	E
£22.50	£2.25	£3.25	£2.50	£2.15

Questions continue on next page

(5) Work out $\frac{3}{5}$ of 145

A	B	C	D	E
29	87	58	78	90

(6) A group of 32 children were asked what pet they liked best.

The pie chart shows the results.

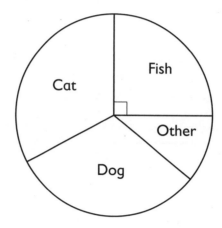

How many of the children liked fish best?

A	B	C	D	E
8	12	16	24	6

(7) The maximum temperatures in the Arctic Circle on seven different days in January were:

−10°C   −23°C   −4°C   2°C   −6°C   −7°C   −14°C

What is the range of the temperatures?

A	B	C	D	E
19°C	4°C	24°C	17°C	25°C

(8) What is the name of the part of the circle the arrow is pointing to?

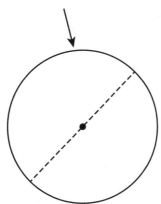

A	B	C	D	E
Radius	Diameter	Circumference	Tangent	Edge

(9) Which of these expressions is equivalent to $2(x - 7) + 2$?

A	B	C	D	E
$(2x - 7) + 2$	$2x - 16$	$2x - 5$	$2x - 12$	$2x - 14$

(10) On a map, 12 mm represents 3.6 m.

What is the scale of the map?

A	B	C	D	E
1 : 30	1 : 3	1 : 300	1 : 3000	3 : 100

Score: .... / 10

Test 15

You have **10 minutes** to complete this test.

You have **10 questions** to complete within the given time.

Circle the letter above the correct answer.

① What is 807 259 rounded to the nearest hundred?

A	B	C	D	E
900 000	807 000	807 200	807 300	807 260

② Gaby books a holiday which costs £370.

She pays a deposit of £45 and then pays the rest of the cost in equal monthly payments over a period of five months.

How much is each monthly payment?

A	B	C	D	E
£65	£70	£60	£74	£83

③ There are 1080 seats on a train.

There are 20 carriages on the train.

How many seats are there in each carriage?

A	B	C	D	E
50	64	44	52	54

④ Hari has 28 marbles.

He gives 7 marbles to Ben.

What fraction of the 28 marbles does Hari have now?

A	B	C	D	E
$\frac{1}{4}$	$\frac{3}{4}$	$\frac{2}{3}$	$\frac{1}{3}$	$\frac{3}{5}$

(5) Here is a bar chart showing the maximum daily temperature in the UK from January to May one year.

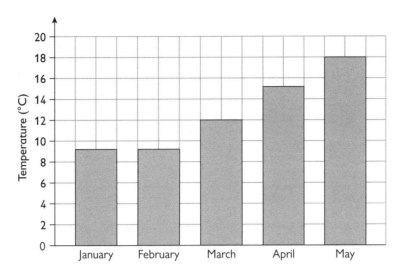

What is the range of maximum daily temperatures?

A	B	C	D	E
4°C	5°C	6°C	9°C	10°C

(6) Here are the first three numbers in a sequence:

3   7   11

What is the sixth number in the sequence?

A	B	C	D	E
4	15	19	23	27

Questions continue on next page

(7) Look at the angle.

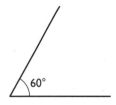

60°

How many of these angles would fit around a point?

A	B	C	D	E
2	3	4	5	6

(8) A television is reduced by 20% in a sale.

The sale price of the television is £200.

What was the price of the television before the sale?

A	B	C	D	E
£240	£250	£220	£180	£210

(9) Write the ratio 125 : 70 in its simplest terms.

A	B	C	D	E
25 : 7	5 : 7	25 : 14	5 : 4	14 : 25

(10) Lily has some marbles.

Alex has twice as many marbles as Lily.

In total, they have 57 marbles.

How many marbles does Alex have?

A	B	C	D	E
19	29	28	31	38

Score: .... / 10

Test 16

You have 10 minutes to complete this test.

You have 10 questions to complete within the given time.

Circle the letter above the correct answer.

① 45 × 57 = 2565

What is 4.5 × 5.7?

A	B	C	D	E
2565	256.5	25.65	2.565	0.2565

② Ravi receives £4.50 pocket money each week.

He spends $\frac{2}{5}$ of his pocket money each week and saves the rest.

For how many weeks must he save up to buy a game costing £45?

A	B	C	D	E
14 weeks	15 weeks	16 weeks	17 weeks	18 weeks

③ What number should go in the box to make these fractions equivalent?

$$\frac{\square}{11} = \frac{44}{121}$$

A	B	C	D	E
4	11	8	2	6

④ A rectangular field is 80 metres long and 35 metres wide.

What is the perimeter of the field?

A	B	C	D	E
115 metres	160 metres	230 metres	195 metres	210 metres

Questions continue on next page

5. Charlie delivers pizzas.

He earns 85p for each pizza he delivers.

One day Charlie earns £30.60

How many pizzas did he deliver on this day?

A	B	C	D	E
36	35	34	32	31

6. What is the missing number in this sequence?

561   554   547   ☐   533

A	B	C	D	E
541	539	538	540	542

7. Here is a shape made of two rectangles.

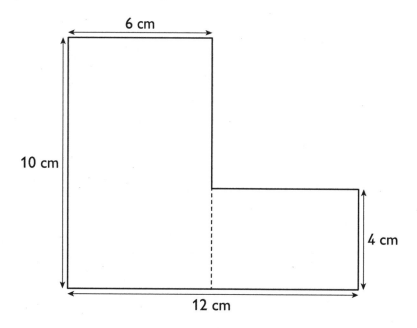

What is the total area of the shape?

A	B	C	D	E
60 cm²	120 cm²	108 cm²	144 cm²	84 cm²

(8) *ABC* is a straight line.

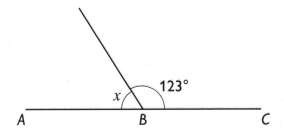

123°

x

A　　　*B*　　　*C*

Work out the size of angle *x*.

A	B	C	D	E
47°	57°	67°	237°	147°

(9) Put these fractions in order of size, starting with the smallest.

$$\frac{1}{3} \qquad \frac{5}{6} \qquad \frac{3}{8} \qquad \frac{7}{12} \qquad \frac{1}{2}$$

A	B	C	D	E
$\frac{1}{2}$ $\frac{1}{3}$ $\frac{3}{8}$ $\frac{5}{6}$ $\frac{7}{12}$	$\frac{1}{3}$ $\frac{1}{2}$ $\frac{3}{8}$ $\frac{7}{12}$ $\frac{5}{6}$	$\frac{1}{2}$ $\frac{3}{8}$ $\frac{1}{3}$ $\frac{7}{12}$ $\frac{5}{6}$	$\frac{1}{3}$ $\frac{3}{8}$ $\frac{1}{2}$ $\frac{7}{12}$ $\frac{5}{6}$	$\frac{5}{6}$ $\frac{7}{12}$ $\frac{1}{2}$ $\frac{3}{8}$ $\frac{1}{3}$

(10) The sum of three consecutive numbers is 120.

What is the largest number?

A	B	C	D	E
41	38	39	40	42

Test 17

You have 10 minutes to complete this test.

You have 10 questions to complete within the given time.

Circle the letter above the correct answer.

(1) Work out 598 ÷ 13

A	B	C	D	E
36	46	38	56	44

(2) Jem makes 24 pizzas.

He sells each pizza for £3.75

Work out the total amount he sells the pizzas for.

A	B	C	D	E
£75	£90	£82.50	£52.50	£55

(3) If $\frac{3}{4}$ of a number is 48, what is the number?

A	B	C	D	E
64	60	36	144	72

(4) The scale drawing below shows a garage and a house.

The real height of the garage is 2.6 metres.

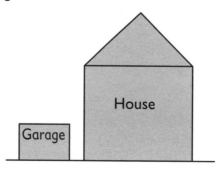

Estimate the real height of the house in metres.

A	B	C	D	E
15.6 metres	10.4 metres	7.8 metres	9 metres	12 metres

(5) What is 5^3?

A	B	C	D	E
15	25	125	555	625

(6) Here is a Venn diagram:

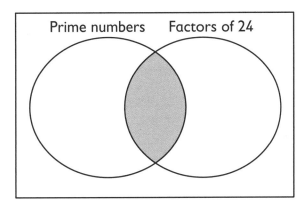

Which numbers go in the shaded section?

A	B	C	D	E
1, 2 and 3	6 and 12	3 and 5	3 and 7	2 and 3

(7) Sarah goes on a bike ride.

She rides at 15 kilometres per hour for 3 hours.

How far does Sarah ride?

A	B	C	D	E
30 km	40 km	3 km	45 km	4.5 km

(8) Five packs of pens are available in a supermarket.

| 7 pens for £1.89 | 5 pens for £1.65 | 2 pens for 62p | 3 pens for 75p | 10 pens for £3 |

Which pack offers the lowest price per pen?

A	B	C	D	E
7 pens for £1.89	5 pens for £1.65	2 pens for 62p	3 pens for 75p	10 pens for £3

Questions continue on next page

(9) The diagram shows a rectangle.

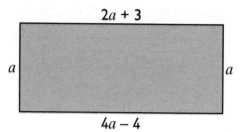

2a + 3

a   a

4a − 4

Work out the value of a.

A	B	C	D	E
2.5	3	3.5	4	4.5

(10) To make purple paint, you can mix red, white and blue paint in the ratio 5 : 2 : 4

How much blue paint is needed to make 77 litres of purple paint?

A	B	C	D	E
35 litres	14 litres	28 litres	44 litres	55 litres

Score: .... / 10

Test 18

You have 10 minutes to complete this test.

You have 10 questions to complete within the given time.

Circle the letter above the correct answer.

1 Work out the value of 0.1 × 0.1

A	B	C	D	E
1	0.1	0.01	0.001	0.0001

2 450 g of fish costs £4.80

Jill buys 300 g of fish.

How much does Jill pay for the fish?

A	B	C	D	E
£2.40	£2.80	£3.20	£3.40	£3.60

3 The area of a square field is 64 square metres.

What is the perimeter of the field?

A	B	C	D	E
8 metres	16 metres	24 metres	32 metres	64 metres

4 What is this number in figures?

Eight hundred and three thousand, and seven

A	B	C	D	E
803 007	80 307	837	8037	800 307

Questions continue on next page

(5) Here is a list of nine numbers:

| 3 | 4 | 4 | 4 | 5 | 5 | 6 | 7 | 7 |

Work out the range of the numbers.

A	B	C	D	E
3	4	5	6	7

(6) What is $\frac{4}{5}$ as a percentage?

A	B	C	D	E
40%	4%	20%	60%	80%

(7) Two numbers are multiplied as shown.

$P7 \times 4 = 348$

What is the value of digit P?

A	B	C	D	E
8	7	6	4	3

(8) The diagram shows the plan of a bedroom.

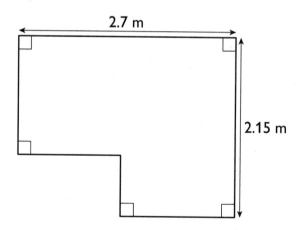

Work out the perimeter of the bedroom.

A	B	C	D	E
4.85 m	14.55 m	6.2 m	9.7 m	10.25 m

(9) An empty container has a mass of 5 kg.

The container is filled with 20 boxes each of mass 750 g.

What is the mass of the container when it has the 20 boxes in it?

A	B	C	D	E
10 kg	15 kg	20 kg	25 kg	30 kg

(10) Which statement is **not true** for an equilateral triangle?

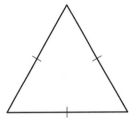

A	B	C	D	E
There are three equal sides	There are three equal angles	Each angle equals 45°	There are three lines of symmetry	There are no parallel sides

Test 19

You have 10 minutes to complete this test.

You have 10 questions to complete within the given time.

Circle the letter above the correct answer.

(1) The diagram below shows the positions of the corners of a square.

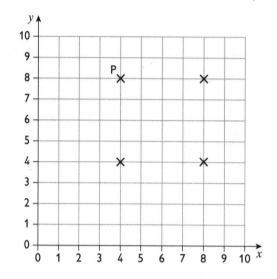

What are the coordinates of the corner marked P?

A	B	C	D	E
(8, 4)	(4, 4)	(8, 8)	(9, 4)	(4, 8)

(2) Which number is **not** a factor of 24?

A	B	C	D	E
1	2	3	9	24

(3) Work out the size of angle x.

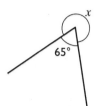

A	B	C	D	E
295°	115°	305°	285°	25°

(4) Here is a sequence:

| −13 | | −5 | −1 | |

What are the two missing numbers?

A	B	C	D	E
−10 and 2	−8 and 4	−9 and 3	−11 and 1	−10 and 3

(5) What is $\frac{1}{2}$ of $\frac{1}{2}$ of $\frac{1}{2}$?

A	B	C	D	E
$\frac{1}{4}$	$\frac{1}{8}$	$\frac{1}{6}$	$\frac{1}{16}$	$\frac{1}{3}$

(6) What is the sum of 2.3 m + 23 cm + 23 mm?

A	B	C	D	E
255.3 cm	48.3 cm	6.9 m	483 mm	2.76 m

(7) What fraction of the whole shape is shaded?

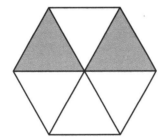

A	B	C	D	E
$\frac{1}{2}$	$\frac{1}{3}$	$\frac{1}{4}$	$\frac{2}{5}$	$\frac{2}{7}$

Questions continue on next page

(8) The graph shows the amount of fuel used by a car on a 250-mile journey.

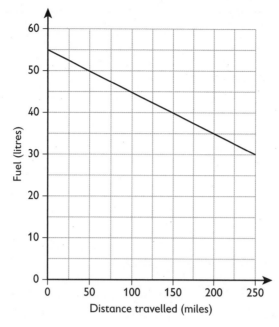

How many more miles can the car travel before it runs out of fuel?

A	B	C	D	E
100	150	200	250	300

(9) Luke is thinking of a number.

The number is 3 more than a prime number.

The number is twice a different prime number.

The digits of the number add up to 8.

What is the number?

A	B	C	D	E
17	35	26	53	44

(10) What is the cube root of 27?

A	B	C	D	E
9	3	7	2	$\frac{1}{3}$

Score: .... / 10

Test 20

You have 10 minutes to complete this test.

You have 10 questions to complete within the given time.

Circle the letter above the correct answer.

1 What is the number forty-seven million, fifty-six thousand and twenty in figures?

A	B	C	D	E
4 705 620	47 560 020	4 756 020	47 056 020	475 620

2 If 6 lots of X equal 72, what do 5 lots of X equal?

A	B	C	D	E
30	35	65	60	45

3 What is the value of 3^4?

A	B	C	D	E
12	81	27	243	7

4 A jug contains 3.5 litres of water.

Ziggy fills 7 cups from the jug.

Each cup holds 300 millilitres.

How much water is left in the jug?

A	B	C	D	E
14 millilitres	14 000 millilitres	1.4 millilitres	1400 millilitres	140 millilitres

Questions continue on next page

(5) How many small squares will fit into the large square?

1 cm   7 cm

A	B	C	D	E
7	21	35	42	49

(6) $-1 \leqslant x < 3$

Which of the following is **not** a possible value of x?

A	B	C	D	E
3	2	1	0	−1

(7) Anju and Raj share some sweets in the ratio 2 : 5

Raj gets 9 more sweets than Anju.

How many sweets does Raj get?

A	B	C	D	E
6	15	24	21	14

(8) The graph shows the mass of a puppy in the first five weeks of its life.

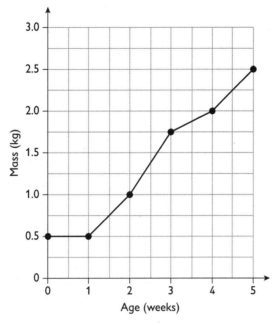

How old was the puppy at the end of the week in which it gained the most mass?

A	B	C	D	E
1 week old	2 weeks old	3 weeks old	4 weeks old	5 weeks old

(9) What is 0.25% of 16 000?

A	B	C	D	E
400	4	4000	0.4	40

(10) $5x - 7 = 2x + 2$

What is the value of x?

A	B	C	D	E
3	2	5	4	1

Test 21

You have 10 minutes to complete this test.

You have 10 questions to complete within the given time.

Circle the letter above the correct answer.

① Which of these has the same value as 6^2?

A	B	C	D	E
$2^2 \times 3$	2×3^2	$2^2 \times 3^2$	2×3	$2^3 \times 3$

② What is the largest prime number less than 50?

A	B	C	D	E
49	47	45	43	41

③ The bar chart shows the number of ice creams sold in two shops one week.

One shop is at a beach and the other is in a town.

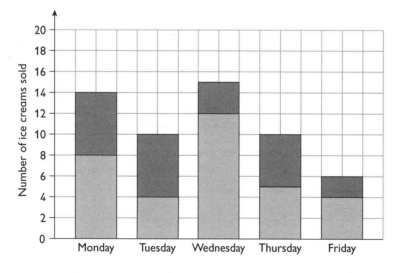

What is the difference between the total number of ice creams sold at the two shops?

A	B	C	D	E
9	10	11	12	13

(4) A map has a scale of 1 : 10 000

The distance between two points on the map is 5 cm.

Work out the real distance between the two points.

A	B	C	D	E
5 km	50 km	0.5 km	500 km	5000 km

(5) The diagram shows a regular pentagon.

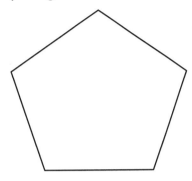

How many lines of symmetry does the regular pentagon have?

A	B	C	D	E
1	2	3	4	5

(6) One of the angles in this isosceles triangle is 65°.

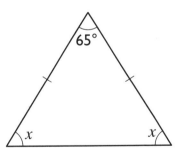

What is the size of each angle labelled x?

A	B	C	D	E
65°	147.5°	12.5°	57.5°	50°

Questions continue on next page

7. The temperatures on five different days in Lapland were:

   −6°C   −5°C   −8°C   1°C   −2°C

   What is the mean of the temperatures?

A	B	C	D	E
−4°C	−8°C	−5°C	9°C	−3°C

8. Sina has 6 green buttons, 7 yellow buttons, 8 red buttons and 9 blue buttons.

   What percentage of her buttons are green?

A	B	C	D	E
20%	25%	40%	6%	18%

9. To make eight glasses of squash, Ali needs 320 ml of orange cordial.

   How many millilitres of orange cordial does he need to make five glasses of squash?

A	B	C	D	E
160 ml	280 ml	240 ml	200 ml	100 ml

10. Ade sits five exams.

    Each exam is out of 100.

    After the first four exams, his mean mark is 78.

    What mark does he need on the last exam to raise his mean mark to 80?

A	B	C	D	E
78	88	98	86	96

Test 22

You have 10 minutes to complete this test.

You have 10 questions to complete within the given time.

Circle the letter above the correct answer.

(1) An aeroplane has 54 rows of 9 seats.

How many seats does the aeroplane have in total?

A	B	C	D	E
6	450	86	477	486

(2) Which number lies exactly halfway between 67.8 and 56.3?

A	B	C	D	E
62	60.2	60.05	62.05	61.2

(3) $x = -2$

What is the value of $3x^2$?

A	B	C	D	E
6	−12	12	−36	36

(4) The probability Ravi's team wins a football match is 0.45

The probability his team loses a football match is 0.2

What is the probability that Ravi's team draws a football match?

A	B	C	D	E
0.53	0.65	0.45	0.4	0.35

Questions continue on next page

(5) What is the highest common factor of 24 and 36?

A	B	C	D	E
2	3	8	12	24

(6) The table below shows information about the colours of cars in a car park.

Colour	Frequency
Red	5
Blue	15
White	16
Silver	24

Daria creates a pie chart using the table.

What angle in the pie chart represents white cars?

A	B	C	D	E
96°	90°	16°	32°	80°

(7) Three equilateral triangles each have a perimeter of 51 cm.

The triangles are joined together to create a trapezium, as shown below.

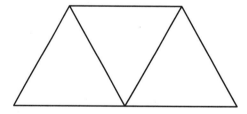

What is the perimeter of the trapezium?

A	B	C	D	E
153 cm	102 cm	119 cm	68 cm	85 cm

(8) Here is a list of ingredients for making 12 biscuits:

Ingredients for 12 biscuits

200 g butter

180 g sugar

1 egg

400 g plain flour

Katya wants to make 30 biscuits.

How much sugar will she need?

A	B	C	D	E
500 g	1000 g	540 g	450 g	600 g

(9) Apples cost 55p each.

Bananas cost 30p each.

The total cost of a apples and b bananas is c.

Find the correct formula for the total cost, in pence, of a apples and b bananas.

A	B	C	D	E
$c = 55 + 30$	$c = 55a + 30b$	$c = a + b$	$c = 30a + 55b$	$c = 55 + a + 30 + b$

(10) A jar contains 5p coins and 10p coins.

The ratio of 5p coins to 10p coins is 5 : 9

There is a total of £3.45 in the jar.

How many 10p coins are there?

A	B	C	D	E
27	135	25	15	18

Score: .... / 10

Notes

Answers

Key abbreviations: °C: degrees centigrade, cm: centimetre, g: gram, kg: kilogram, km: kilometre, m: metre, ml: millilitre, mm: millimetre

Test 1

Q1 **B**
3 hundredths

Q2 **E**
18 × 5 = 50 + 40 = 90

Q3 **A**
22:45 to 08:45 is 10 hours
+ 15 minutes is 09:00
+ 37 minutes is 09:37
15 + 37 = 52
The total time is 10 hours 52 minutes

Q4 **B**
−3 − 8 = −11
−11 + 6 = −5

Q5 **D**
£11.25 ÷ 5 = £2.25 (cost of one cake)
Three cakes cost 3 × £2.25 = £6.75
£13.75 − £6.75 = £7.00
Four teas cost £7.00
One tea costs £7.00 ÷ 4 = £1.75

Q6 **A**
200 mm = 20 cm, 1.2 m = 120 cm, 1.1 m = 110 cm
120 + 110 + 67 = 297 cm
297 ÷ 3 = 99 cm

Q7 **E**
$x + 2x = 90°$
$3x = 90°$
$x = 30°$

Q8 **B**
The next pattern has five dots on the bottom row.
So fifth pattern dots = 10 + 5 = 15
The sixth pattern has an extra six dots = 15 + 6 = 21

Q9 **C**
Adding up the other four bars gives
160 + 110 + 30 + 80 = £380

Q10 **A**
Let the number be n.
$2n + 13 = 27$
Subtracting 13 from both sides gives $2n = 14$
Then dividing both sides by 2 gives $n = 7$

Test 2

Q1 **B**
128 = 64 × 2
So 15 168 × 2 gives the answer 30 336

Q2 **A**
$\frac{1}{4}$ of 24 = 6 (blue marbles)
24 − 6 = 18 (green marbles)

Q3 **E**
13 − 20 = −7

Q4 **D**
The squares have side length of 12 ÷ 3 = 4 cm
There are 14 sides around the perimeter of the shape.
The perimeter is 14 × 4 = 56 cm

Q5 **A**
Shape A has 5 sides.

Q6 **E**
The difference between the terms increases by
1 each time.
13 + 5 = 18

Q7 **C**
7 × 83p = 581p = £5.81
£10.00 − £5.81 = £4.19

Q8 **A**
6 000 000 − 45 000 = 5 955 000

Q9 **A**
$5 × n = 30$
$n = 6$
$3 × 6 = 18$

Q10 **D**
The 25 lamp posts will have 24 spaces between them.
840 ÷ 24 = 35 m

Test 3

Q1 **C**
423 × 14 gives the answer 5922

Q2 **E**
36 ÷ 4 = 9

Q3 **D**
19 + 31 + 31 + 27 + 26 + 19 = 153 minutes

Q4 **A**
Each side of the room measures 3 m
Split the room into two parts (by drawing a vertical line).
One part is 3 × 2 = 6 m² and the other part is 1 × 0.8 = 0.8 m²
The total area is 6 + 0.8 = 6.8 m²

Q5 **A**
3.5 litres = 3500 ml
$\frac{1}{5}$ of 3500 = 700 ml
3500 − 700 − 650 = 2150 ml

Q6 **B**
£850 − £275 (deposit) = £575
£575 ÷ 23 monthly payments = £25

Q7 **C**
0.0473

Q8 **E**
(4 × 10) − 8 = 32
$x = 10$

Q9 **D**

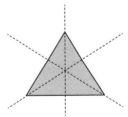

3 lines of symmetry

Q10 **B**

$$\frac{1}{2} + \frac{2}{3} + \frac{3}{4} = \frac{6}{12} + \frac{8}{12} + \frac{9}{12} = \frac{23}{12}$$

$$\frac{23}{12} \div 3 = \frac{23}{36}$$

Test 4

Q1 **B**

9 hundred thousand

Q2 **B**

24 + 12 + 20 + 10 = 66

Q3 **D**

$$\frac{12}{100} = \frac{6}{50} = \frac{3}{25}$$

Q4 **E**

5

Q5 **C**

26 + a (daughter's age now) = 3 × a

26 = 2a

a = 13

Q6 **B**

7.56 pm + 1 hour = 8.56 pm

8.56 pm + 53 minutes = 9.49 pm

9.49 pm + 19 minutes = 10.08 pm

Q7 **C**

$b = 5a - 7$

Q8 **A**

There are five parts altogether.

25 ÷ 5 = 5 litres per part

Two parts are red.

2 × 5 = 10 litres

Q9 **A**

Four small cubes can fit along each side of the large cube.

4 × 4 × 4 = 64

Q10 **B**

Test 5

Q1 **C**

405 027

Q2 **B**

215 students join, so 1037 + 215 = 1252

178 students leave, so 1252 − 178 = 1074

Q3 **D**

93 + 29 = 122 cm

122 cm = 1.22 m

Q4 **E**

There are 180° in a triangle.

The total of $x + y + z$ is 180°

Angle $x = 180° - (y + z)$

Q5 **D**

(4 × number) − 9 = 27

4 × number = 27 + 9 = 36

number = 36 ÷ 4 = 9

Q6 **B**

2.73 ÷ 0.01 = 273

Q7 **A**

The total journey takes 66 + 23 + 54 minutes = 143 minutes = 2 hours 23 minutes

The journey starts at 08:49

Adding 2 hours 23 minutes takes you to 11:12

Q8 **D**

The area of a triangle is found by multiplying $\frac{1}{2}$ × base × height.

$56 = \frac{1}{2} \times 8 \times$ height

= 4 × height, so height of the triangle is 14 cm.

Q9 **D**

Net D is the only one which cannot be folded into a cube.

Q10 **B**

The largest prime number between 20 and 30 is 29

The smallest one is 23

The difference is 29 − 23 = 6

Test 6

Q1 **D**

19 × 5 = 95

Q2 **A**

If 26 × 18 = 468, then 468 ÷ 18 = 26

46.8 ÷ 18 = 2.6 and so 46.8 ÷ 1.8 = 26

Q3 **B**

The difference between 731 and 714 = 17

The difference between 714 and 697 = 17

The missing number is 731 + 17 = 748

To check, you can calculate 765 − 17 = 748

Q4 **E**

On Monday he ran 11 km, Tuesday 6 km, Wednesday 10 km, Thursday 15 km and Friday 8 km.

11 + 6 + 10 + 15 + 8 = 50 km

Q5 **A**

Using the order of operations, apply the index number and then do the multiplication.

$4^2 = 16$ and 3 × 5 = 15

Finally, do the subtraction, 16 − 15 = 1

Q6 **B**

$$\frac{1}{12} + \frac{1}{4} = \frac{1}{12} + \frac{3}{12} = \frac{4}{12} = \frac{1}{3}$$

$$\frac{1}{3} \div 2 = \frac{1}{3} \times \frac{1}{2} = \frac{1}{6}$$

$$\frac{1}{3} - \frac{1}{6} = \frac{2}{6} - \frac{1}{6} = \frac{1}{6}$$

Q7 **B**

Point A is translated 1 unit to the right and 5 units up.

Point C has coordinates (6, 1).

When translated, it moves 1 unit right and 5 units up to (6 + 1, 1 + 5) = (7, 6)

Q8 **A**

$\frac{5}{8}$ of 240 = 240 ÷ 8 × 5 = 30 × 5 = 150

To find 40% of 150, first find 10% of 150 = 15

Then multiply by 4 to find 40% of 150 = 60

Q9   **D**

If five coffees cost £11.75 then one coffee costs

11.75 ÷ 5 = £2.35

Three coffees cost 3 × 2.35 = £7.05

Four teas must cost 15.45 − 7.05 = £8.40

One tea costs 8.40 ÷ 4 = £2.10

Q10   **E**

There are 360 ÷ 12 = 30° between each hour number
on a clock.

At 2:30, the minute hand is at the number 6 and the
hour hand is halfway between 2 and 3, which is 15°.

There are 90° between the numbers 3 and 6.

The total angle is 15 + 90 = 105°

Test 7

Q1   **D**

67.9

Q2   **B**

£9.50 × 8 = £76

£76 × 5 = £380

£380 × 3 = £1140

Q3   **E**

4662

Q4   **B**

1 000 000 mm ÷ 10 = 100 000 cm

100 000 cm ÷ 100 = 1000 m

1000 m ÷ 1000 = 1 km

Q5   **A**

nth term is $7n − 3$

(7 × 11) − 3 = 74

Q6   **C**

24 ÷ 3 = £8 per part

£8 × 5 = £40

Q7   **E**

10% of 450 = 45

5% of 450 = 22.5

1% of 450 = 4.5

45 + 22.5 + 4.5 = 72

Q8   **D**

Turning the smaller rectangle 90°, it will fit 4 across
and 5 down the larger rectangle.

4 ÷ 1 = 4 and 15 ÷ 3 = 5

4 × 5 = 20

Q9   **E**

There is only one pair of parallel sides is not true.

Q10   **B**

45 ÷ 3 = 15 (middle of the three consecutive
odd numbers)

13 + 15 + 17 = 45

13

Test 8

Q1   **A**

$\frac{29}{10}$

Q2   **B**

4 + 6 + 2 + 10 + 16 = 38

Q3   **C**

10 − 2 = 8

8 × 180° = 1440°

Q4   **D**

(1, 4) and (7, 10)

Q5   **A**

$\frac{1}{2}$ × 1.2 × 0.8

0.48 cm²

Q6   **D**

20% of £125 = £25

£125 − £25 = £100

5% of £100 = £5

£100 − £5 = £95

Q7   **B**

56 ÷ 8 = 7

7 × 5 = 35 miles (not 30)

56 kilometres = 30 miles is the wrong answer.

Q8   **B**

There are three equal sides in an equilateral triangle.

22.5 ÷ 3 = 7.5 cm

Q9   **E**

2 × £2.65 = £5.30

3 × £2.45 = £7.35

£5.30 + £7.35 + £2.95 = £15.60

Q10   **D**

$2^3 × 2^2 × 2 = 8 × 4 × 2 = 64$

$2^3 × 2^3 = 8 × 8 = 64$

$4^3 = 4 × 4 × 4 = 64$

$4^2 × 4 = 16 × 4 = 64$

$2^3 × 4 = 8 × 4 = 32$

$8^2 = 8 × 8 = 64$

$2^3 × 4$ does not equal $2^3 × 2^2 × 2$

Test 9

Q1   **C**

$(2 × −4) − (−2 × 2) = −8 − −4$

$ = −8 + 4$

$ = −4$

Q2   **B**

73 × 100 = 7300

73 × 99 = 7300 − 73

73 × 98 = 7300 − 146

Q3   **A**

8.45 am + 2 hours 45 minutes ($2\frac{3}{4}$ hours) is 11.30 am

11.30 am + 20 minutes is 11.50 am

11.50 am + 1 hour 35 minutes (95 minutes) is 1.25 pm

Q4   **B**

28.5 ÷ 3 = 9.5

$x + 7 = 9.5$

$x = 2.5$

Q5   **E**

7

Q6   **D**

£5 = €6 and so £10 = €12

£2.50 = €3

£17.50 = €6 + €12 + €3 = €21

Q7   **B**

4 litres = 4000 ml

4000 ÷ 250 = 16

Q8 **D**

There are 180° in a triangle.

Equilateral triangles have three equal angles.

180 ÷ 3 = 60°

Q9 **E**

365 ÷ 6 = 60 remainder 5, so 60 boxes were filled

Q10 **A**

$\frac{1}{4} + \frac{3}{5} = \frac{5}{20} + \frac{12}{20} = \frac{17}{20}$

$1 - \frac{17}{20} = \frac{3}{20}$

$\frac{3}{20} = 87$ and so $\frac{1}{20} = 87 \div 3 = 29$

$29 \times 20 = 580$

Test 10

Q1 **C**

0.03

Q2 **D**

£1.65 × 12 = £19.80

£19.80 − £15 = £4.80

Q3 **E**

3 + 10 = 13 minutes

Q4 **E**

There are 13 segments equalling 26 hours.

26 ÷ 13 = 2 hours per segment

Friday has two segments, equalling 4 hours of sunshine.

Q5 **C**

$2x + 1 + 3x - 2 + x = 6x - 1$

(6 × 2) − 1 = 11 cm

Q6 **A**

If $-3 < x < 2$, x cannot equal −3 as it is greater than −3

Q7 **B**

C to G

Q8 **A**

£7 = 700p

$\frac{35}{700} = \frac{5}{100}$

5%

Q9 **D**

1 − 0.43 = 0.57

Q10 **C**

Let my age now be a.

$4a - 5 = 5(a - 5)$

$4a - 5 = 5a - 25$

$4a + 20 = 5a$

$a = 20$

Testing by using numbers:

4 × 20 = 80

80 − 5 = 75 and 20 − 5 = 15

15 × 5 = 75

Test 11

Q1 **B**

7.2 ÷ ? = 720 and so 7.2 ÷ 720 = ?

Multiplying both values by 10 gives 72 ÷ 7200 = 0.01

Q2 **B**

20 cm   24 mm = 2.4 cm   1.2 m = 120 cm

0.1 m = 10 cm   0.03 m = 3 cm

3 cm − 2.4 cm = 0.6 cm = 6 mm

Q3 **E**

2, 3, 5, 7, 11, 13

Q4 **C**

£625.20 ÷ 24 = £26.05

Q5 **A**

237 ÷ 3 = 79p per pen

112 ÷ 2 = 56p per ruler

(79 × 10) + (56 × 5) = 1070p = £10.70 in total

Q6 **E**

750 × 3 = 2250 g

3500 − 2250 = 1250 g

1250 ÷ 250 = 5 small tins

Q7 **D**

Angle $x > 180°$

Q8 **D**

5 × 6 = 30

2 + 6 + 7 + 9 + ? = 30

24 + ? = 30

The missing card is 6

Q9 **A**

6.098 + 0.007 = 6.105

Q10 **A**

$3x - 1 + 4x + 3 + 3x - 1 + 4x + 3 = 14x + 4 = 2(7x + 2)$

$= 7x + 2 + 7x + 2 = 2(3x - 1) + 2(4x + 3)$

The expression that does not represent the perimeter is $7x + 2$.

Test 12

Q1 **C**

50 × 4 = 200, 7 × 4 = 28, 200 + 28 = 228

Q2 **D**

$\frac{2}{3} - \frac{1}{4} = \frac{8}{12} - \frac{3}{12} = \frac{5}{12}$

Q3 **C**

There are 1000 metres in 1 kilometre.

13 457 ÷ 1000 = 13.457

Q4 **A**

1

Q5 **E**

16

Q6 **D**

1% of £70 = 70p

7% of £70 = 70p × 7 = 490p = £4.90

Q7 **B**

$\frac{3}{4} \times 2 \times 7 = 10.5$

So she will need 11 tins

Q8 **C**

25

Q9 **B**

A sphere does not have two identical flat faces (ends) and uniform cross-section across its length.

Q10 **C**

Beth's age is $x + 2$ so Marie's age is $2(x + 2)$.

Test 13

Q1 **B**

30 000 × 100 = 3 000 000

Q2 **A**

141 − 27 = 114

114 ÷ 2 = 57

Ben saved £57 (Abby saved £57 + £27 = £84)

Q3   **D**
500 ÷ 47 = 10 remainder 30p

Q4   **E**
$\frac{3}{4} \times \frac{1}{3} = \frac{3}{12} = \frac{1}{4}$

Q5   **B**
103 − 17 = 86 cm = 0.86 m

Q6   **C**
1.4 kg = 1400 g
560 + 970 + 1400 = 2930 g
2930 ÷ 1000 = 2.93 kg

Q7   **A**
5 × 1.6 = 8 km

Q8   **D**
−6, −1, 4, 9, 14, **19**

Q9   **D**
Jakub cycled to his friend's house, stayed for lunch, and then cycled back.

Q10   **E**
2.7 ÷ 30 = 27 ÷ 300 = 27 ÷ 3 ÷ 100 = 0.09

Test 14

Q1   **C**
3 × 13 = 39
47 − 39 = 8

Q2   **E**
42

Q3   **A**
40% of 140 = 56
25% of 220 = 55
56 − 55 = 1

Q4   **B**
315 ÷ 3.5 = 90p per kilogram
90 × 2.5 = 225p = £2.25

Q5   **B**
145 ÷ 5 = 29
29 × 3 = 87

Q6   **A**
8

Q7   **E**
2 − (−23) = 25
25°C

Q8   **C**
The circumference is the perimeter of a circle.

Q9   **D**
$2(x − 7) + 2 = 2x − 14 + 2$
$2x − 12$

Q10   **C**
3.6 m = 360 cm = 3600 mm
3600 ÷ 12 = 300
The map scale is 1 : 300

Test 15

Q1   **D**
807 300

Q2   **A**
£370 − £45 = £325
£325 ÷ 5 = £65

Q3   **E**
1080 ÷ 20 = 54

Q4   **B**
28 − 7 = 21
$\frac{21}{28} = \frac{3}{4}$

Q5   **D**
18 − 9 = 9°C

Q6   **D**
3, 7, 11, 15, 19, **23**

Q7   **E**
There are 360° around a point.
360 ÷ 60 = 6

Q8   **B**
£200 = 80%
£50 = 20%
Original price = £200 + £50 = £250

Q9   **C**
125 ÷ 5 = 25
70 ÷ 5 = 14
25 : 14

Q10   **E**
57 ÷ 3 = 19
19 × 2 = 38

Test 16

Q1   **C**
25.65

Q2   **D**
$\frac{3}{5}$ of £4.50 = £4.50 ÷ 5 × 3 = £2.70
£45 ÷ £2.70 = 16.666…
17 weeks

Q3   **A**
121 ÷ 11 = 11
44 ÷ 11 = 4

Q4   **C**
80 + 35 + 80 + 35 = 230 metres

Q5   **A**
3060 ÷ 85 = 36

Q6   **D**
The sequence goes down in 7s.
547 − 7 = 540

Q7   **E**
6 × 10 = 60 cm² (area of larger rectangle)
12 − 6 = 6 cm (length of smaller rectangle)
6 × 4 = 24 cm² (area of smaller rectangle)
Total area = 60 + 24 = 84 cm²

Q8   **B**
180° − 123° = 57°

Q9   **D**
$\frac{1}{3} = \frac{8}{24}$ $\frac{3}{8} = \frac{9}{24}$ $\frac{1}{2} = \frac{12}{24}$ $\frac{7}{12} = \frac{14}{24}$ $\frac{5}{6} = \frac{20}{24}$
$\frac{1}{3}$ $\frac{3}{8}$ $\frac{1}{2}$ $\frac{7}{12}$ $\frac{5}{6}$

Q10   **A**
120 ÷ 3 = 40
39 + 40 + 41 = 120
Largest number is 41

Test 17

Q1 **B**
46

Q2 **B**
24 × £3.75 = (20 × £3.75) + (4 × £3.75)
= £75 + £15
= £90

Q3 **A**
48 ÷ 3 × 4 = 64

Q4 **B**
The house is four times the height of the garage.
2.6 × 4 = 10.4 metres

Q5 **C**
5^3 = 5 × 5 × 5 = 125

Q6 **E**
Factors of 24 are 1, 2, 3, 4, 6, 8, 12, 24
Prime numbers which are factors of 24 are 2 and 3

Q7 **D**
15 × 3 = 45 km

Q8 **D**
7 pens for £1.89 = 27p per pen
5 pens for £1.65 = 33p per pen
2 pens for 62p = 31p per pen
3 pens for 75p = 25p per pen   (lowest price)
10 pens for £3 = 30p per pen

Q9 **C**
2a + 3 = 4a − 4
    3 = 2a − 4
    7 = 2a
    a = 3.5

Q10 **C**
5 + 2 + 4 = 11 parts
77 ÷ 11 = 7 litres per part
4 × 7 = 28 litres

Test 18

Q1 **C**
0.01

Q2 **C**
150 g costs £4.80 ÷ 3 = £1.60
300 g costs £1.60 × 2 = £3.20

Q3 **D**
$\sqrt{64}$ = 8
Length of side of the field = 8 m
Perimeter = 8 + 8 + 8 + 8 = 32 m

Q4 **A**
803 007

Q5 **B**
7 − 3 = 4

Q6 **E**
$\frac{4}{5} = \frac{8}{10} = \frac{80}{100}$
80%

Q7 **A**
87 × 4 = 348
P = 8

Q8 **D**
2.7 + 2.7 + 2.15 + 2.15 = 9.7 m

Q9 **C**
750 × 20 = 15 000 g
15 000 g ÷ 1000 = 15 kg
15 + 5 = 20 kg

Q10 **C**
Angles in an equilateral triangle equal 60°, not 45°.

Test 19

Q1 **E**
(4, 8)

Q2 **D**
9

Q3 **A**
360° − 65° = 295°

Q4 **C**
The difference between each term in the sequence is +4
−13 + 4 = −9 and −1 + 4 = 3
The missing numbers are −9 and 3

Q5 **B**
$\frac{1}{2} × \frac{1}{2} × \frac{1}{2} = \frac{1}{8}$

Q6 **A**
2.3 m = 230 cm
23 mm = 2.3 cm
230 + 23 + 2.3 = 255.3 cm

Q7 **B**
$\frac{2}{6} = \frac{1}{3}$

Q8 **E**
25 litres = 250 miles
1 litre = 10 miles
30 litres = 300 miles

Q9 **C**
26

Q10 **B**
3

Test 20

Q1 **D**
47 056 020

Q2 **D**
72 ÷ 6 = 12
5 × 12 = 60

Q3 **B**
3^4 = 3 × 3 × 3 × 3 = 81

Q4 **D**
3.5 litres = 3500 millilitres
7 × 300 = 2100 millilitres
3500 − 2100 = 1400 millilitres

Q5 **E**
7 × 7 = 49

Q6 **A**
3

Q7 **B**
5 − 2 = 3
3 parts = 9 sweets
1 part = 3 sweets
Raj gets 5 parts
5 × 3 = 15

Q8 **C**
The graph is steepest between 2 and 3 weeks.
So 3 weeks old.

Q9   **E**

1% of 16 000 = 160

0.25% is one-quarter of 1%

One-quarter of 160 = 40

Q10   **A**

$5x - 7 = 2x + 2$

$3x - 7 = 2$

$ 3x = 9$

$ x = 3$

Test 21

Q1   **C**

$2^2 \times 3^2$

Q2   **B**

47

Q3   **C**

33 − 22 = 11

Q4   **C**

5 × 10 000 = 50 000 cm

50 000 cm ÷ 100 = 500 m

500 m = 0.5 km

Q5   **E**

5

Q6   **D**

180 − 65 = 115

115 ÷ 2 = 57.5°

Q7   **A**

−6 + −5 + −8 + 1 + −2 = −20

−20 ÷ 5 = −4

Q8   **A**

$\frac{6}{30}$ = 20%

Q9   **D**

320 ÷ 8 = 40 ml per glass

40 × 5 = 200 ml

Q10   **B**

4 × 78 = 312 marks after four exams

5 × 80 = 400 marks needed in total

400 − 312 = 88 marks needed in the fifth exam

Test 22

Q1   **E**

54 × 9 = 486

Q2   **D**

67.8 + 56.3 = 124.1

124.1 ÷ 2 = 62.05

Q3   **C**

$3 \times (-2)^2 = 3 \times 4 = 12$

Q4   **E**

0.45 + 0.2 = 0.65

1 − 0.65 = 0.35

Q5   **D**

12

Q6   **A**

5 + 15 + 16 + 24 = 60 cars in total

360° in a circle divided by 60 = 6

This means each car is represented by 6° in the pie chart.

16 white cars × 6 = 96°

Q7   **E**

51 ÷ 3 = 17 cm side length on each equilateral triangle.

The trapezium has 5 lots of 17 cm on its perimeter.

5 × 17 = 85 cm

Q8   **D**

180 g of sugar makes 12 biscuits

90 g of sugar makes 6 biscuits (divide by 2)

450 g of sugar makes 30 biscuits (multiply by 5)

Q9   **B**

$c = 55a + 30b$

Q10   **A**

5 × 5 = 25 and 9 × 10 = 90

25 + 90 = 115

345 ÷ 115 = 3 lots of the ratio 5 : 9

3 × 9 (10p coins) = 27

Notes

Notes

Notes